LIZARDS
OF BORNEO

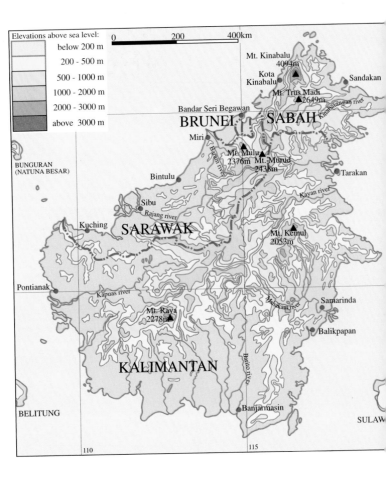

Map of Borneo.

LIZARDS
OF BORNEO

Text and photographs by

Indraneil Das

Natural History Publications (Borneo)
Kota Kinabalu

2004

Published by

Natural History Publications (Borneo) Sdn. Bhd. (216807-X)
A913, 9th Floor, Wisma Merdeka
P.O. Box 15566
88864 Kota Kinabalu, Sabah, Malaysia
Tel: 088-233098 Fax: 088-240768
e-mail: chewlun@tm.net.my
website: www.nhpborneo.com

First published January 2004

Lizards of Borneo by Indraneil Das

Design and layout by Cheng Jen Wai

ISBN 983-812-080-4

Frontispiece: Map of Borneo.

Printed in Malaysia

Contents

Introduction

Borneo, one of the major islands of the Indo-Malayan Archipelago, is the second largest tropical island in the world (after New Guinea), covering a land area of approximately 743,380 sq km. A major part of the island constitutes the Indonesian possession of Kalimantan (area: 539,460 sq km), the rest within the east Malaysian states of Sarawak (124,450 sq km) and Sabah (73,710 sq km), and a small but significant portion is the Sultanate of Negara Brunei Darussalam (5,760 sq km).

The island is situated entirely within the tropics, with the equator crossing over the city of Pontianak in Kalimantan. The land is thus characterised by high, equitable temperature and heavy rainfall spread throughout the year, although relatively wetter periods are noticed during the Northeast Monsoons (November to April), rainfall is also experienced during the passage of the Southwest Monsoons (April to August).

Forest type most typical of Borneo must be the mixed dipterocarp type, from the yellow-red soils in the uplands. The lower montane and upper montane forests are also remarkable in their structure and composition. Here, canopy height is reduced, sometimes to 18–30 m, with few emergent trees, buttressed trees are less common, there is an absence of large woody climbers,

---

At 4095 m, Low's Peak on Gunung Kinabalu is the highest mountain peak in Borneo, and the Kinabalu massif itself is an important centre for lizard diversity, with as many as 45 species recorded. (Photo: Tee Kim Ling).

which is offset by the great abundance of vascular epiphytes. Moss vegetation, with an abundance of bryophytes, in addition to gnarled trees characterise the upper limits of montane forests. Upper montane forests also have trees with small, leathery leaves and conifers. Forests associated with peat swamps are particularly distinctive, and are widespread along coastal areas abutting estuarine plains as well as in small river valleys. Plant life is adapted to high-stress environments, including mineral deficient substrate and poorly-aerated and acidic waters, in addition to shortage of surface water during dry periods. Another unique vegetation type is the heath forests or Kerangas, confined to either raised beach terraces or sandstone ridges and plateaux, formed on ancient sea beaches left stranded by the fall in sea levels, about a million years before present.

Distinctive vegetation types also include the local limestone flora, speciose in plant and invertebrate (especially mollusc) endemics. Mangroves forests are also rich in species. The coastal margins are dominated by mangroves, while more freshwater conditions support strands of *Nypa fruticans* palms. Further upriver, declining salinity levels promote the association of *Heritiera littoralis* and *Oncosperma tigillarium*. Mangroves of the Sunda Shelf islands have been described as the most biologically diverse in the world, and trees may reach 50 m in height.

This work is a non-technical guide to the lizards inhabiting Borneo and its offshore islands. At present, 109 species are known from the region (Table 1), and a larger work on the fauna is in preparation. The present guide covers 73 species, or over 66% of the fauna.

For each species, the current valid scientific name, maximum size (snout-vent length) and brief notes on identification, biology and distribution, are provided.

Table 1. Checklist of Bornean lizards. *indicates endemic species.

AGAMIDAE
Aphaniotis **Peters, 1864**
Aphaniotis acutirostris Modigliani, 1889
Aphaniotis fusca (Peters, 1864)
* *Aphaniotis ornata* (van Lidth de Jeude, 1893)
Bronchocela **Kaup, 1827**
Bronchocela cristatella (Kuhl, 1820)
Bronchocela jubata Duméril & Bibron, 1837
Complicitus **Manthey & Grossmann, 1997**
* *Complicitus nigrigularis* (Ota & Hikida, 1991)
Draco **Linnaeus, 1758**
* *Draco affinis* Bartlett, 1895
Draco cornutus Günther, 1864
Draco cristatellus Günther, 1872
Draco fimbriatus Kuhl, 1820
Draco haematopogon Boie in: Gray, 1831
Draco maximus Boulenger, 1893
Draco melanopogon Boulenger, 1887
Draco obscurus Boulenger, 1887
Draco quinquefasciatus Hardwicke & Gray, 1827
Draco sumatranus Schlegel, 1844
Gonocephalus **Kaup, 1825**
* *Gonocephalus bornensis* (Schlegel, 1848)
Gonocephalus doriae (Peters, 1871)
Gonocephalus grandis (Gray, 1845)
Gonocephalus liogaster (Günther, 1872)
* *Gonocephalus mjobergi* Smith, 1925
Harpesaurus **Boulenger, 1885**
* *Harpesaurus borneensis* (Mertens, 1924)
Hypsicalotes **Manthey & Denzer, 2000**
* *Hypsicalotes kinabaluensis* (De Grijs, 1937)

***Phoxophrys* Hubrecht, 1881**
* *Phoxophrys borneensis* Inger, 1960
* *Phoxophrys cephalum* (Mocquard, 1890)
* *Phoxophrys nigrilabris* (Peters, 1864)
* *Phoxophrys spiniceps* Smith, 1925

***Pseudocalotes* Fitzinger, 1843**
* *Pseudocalotes sarawacensis* Inger & Stuebing, 1994

ANGUIDAE
***Ophisaurus* Daudin, 1803**
* *Ophisaurus buettikoferi* van Lidth de Jeude, 1905

EUBLEPHARIDAE
***Aeluroscalabotes* Boulenger, 1885**
Aeluroscalabotes felinus (Günther, 1864)

DIBAMIDAE
***Dibamus* Duméril & Bibron, 1839**
* *Dibamus ingeri* Das & Lim, 2003
Dibamus leucurus (Bleeker, 1860)
* *Dibamus vorisi* Das & Lim, 2003

GEKKONIDAE
***Cnemaspis* Strauch, 1887**
* *Cnemaspis dringi* Das & Bauer, 1998
Cnemaspis kendallii (Gray, 1845)
Cnemaspis nigridia (Smith, 1925)
***Cosymbotus* Fitzinger, 1843**
Cosymbotus craspedotus (Mocquard, 1890)
Cosymbotus platyurus (Schneider, 1792)
***Cyrtodactylus* Hardwicke & Gray, 1827**
* *Cyrtodactylus baluensis* (Mocquard, 1890)
* *Cyrtodactylus cavernicolus* Inger & King, 1961

4

Cyrtodactylus consobrinus (Peters, 1871)
* *Cyrtodactylus ingeri* Hikida, 1990
* *Cyrtodactylus malayanus* (De Rooij, 1915)
* *Cyrtodactylus matsuii* Hikida, 1990
* *Cyrtodactylus pubisulcus* Inger, 1957
Cyrtodactylus quadrivirgatus Taylor, 1962
* *Cyrtodactylus yoshii* Hikida, 1990
Gehyra Gray, 1834
Gehyra mutilata (Wiegmann, 1834)
Gekko Laurenti, 1768
Gekko gecko Linnaeus, 1758
Gekko monarchus (Duméril & Bibron, 1836)
Gekko smithii (Gray, 1842)
Hemidactylus Gray, 1825
Hemidactylus brookii Gray, 1845
Hemidactylus frenatus Duméril & Bibron, 1836
Hemidactylus garnotii Duméril & Bibron, 1836
Hemiphyllodactylus Bleeker 1860
Hemiphyllodactylus typus Bleeker, 1860
Lepidodactylus Fitzinger, 1843
Lepidodactylus lugubris (Duméril & Bibron, 1836)
* *Lepidodactylus ranauensis* Ota & Hikida, 1988
Luperosaurus Gray, 1845
Luperosaurus browni Russell, 1979
* *Luperosaurus yasumai* Ota, Sengoku & Hikida, 1996
Ptychozoon Kuhl & van Hasselt, 1822
Ptychozoon horsfieldii (Gray, 1827)
Ptychozoon kuhli Stejneger, 1902
* *Ptychozoon rhacophorus* (Boulenger, 1899)

LACERTIDAE
Takydromus Daudin, 1802
Takydromus sexlineatus Daudin, 1802

LANTHANOTIDAE
  Lanthanotus **Steindachner, 1877**
* *Lanthanotus borneensis* Steindachner, 1877

SCINCIDAE
  Apterygodon **Edeling, 1864**
* *Apterygodon vittatus* Edeling, 1864
  Brachymeles **Duméril & Bibron, 1839**
* *Brachymeles apus* Hikida, 1982
  Dasia **Gray, 1839**
  Dasia grisea (Gray, 1845)
  Dasia olivacea Gray, 1839
  Dasia semicincta (Peters, 1867)
  Emoia **Gray, 1845**
  Emoia atrocostata (Lesson, 1830)
  Emoia caeruleocauda (De Vis, 1892)
  Emoia cyanura (Lesson, 1830)
  Lamprolepis **Fitzinger, 1843**
* *Lamprolepis nieuwenhuisii* (van Lidth de Jeude, 1905)
* *Lamprolepis vyneri* (Shelford, 1905)
  Larutia **Böhme, 1981**
* *Larutia* undescribed species Grismer, Leong & Yaakob, 2003
  Lipinia **Gray, 1845**
* *Lipinia* undescribed species Das and Austin, in prep
* *Lipinia miangensis* (Werner, 1910)
* *Lipinia nitens* (Peters, 1871)
  Lipinia vittigera (Boulenger, 1894)
  Lygosoma **Hardwicke & Gray, 1827**
  Lygosoma bampfyldei Bartlett, 1895
  Lygosoma bowringii (Günther, 1864)
  Mabuya **Fitzinger, 1826**
  Mabuya indeprensa Brown & Alcala, 1980
  Mabuya multifasciata (Kuhl, 1820)

Mabuya rudis Boulenger, 1887
Mabuya rugifera (Stoliczka, 1870)
Sphenomorphus Fitzinger, 1843
* *Sphenomorphus aesculeticola* Inger, Tan, Lakim & Yambun, 2002
* *Sphenomorphus buettikoferi* (van Lidth de Jeude, 1905)
* *Sphenomorphus crassa* Inger, Tan, Lakim & Yambun, 2002
Sphenomorphus cyanolaemus Inger & Hosmer, 1965
* *Sphenomorphus haasi* Inger & Hosmer, 1965
* *Sphenomorphus hallieri* (van Lidth de Jeude, 1905)
* *Sphenomorphus kinabaluensis* (Bartlett, 1895)
* *Sphenomorphus maculicollus* Bacon, 1967
* *Sphenomorphus multisquamatus* Inger, 1958
* *Sphenomorphus murudensis* Smith, 1925
* *Sphenomorphus sabanus* Inger, 1958
* *Sphenomorphus shelfordi* (Boulenger, 1900)
Sphenomorphus stellatus (Boulenger, 1900)
* *Sphenomorphus tanahtinggi* Inger, Tan, Lakim & Yambun, 2002
* *Sphenomorphus tenuiculus* (Mocquard, 1890)
Tropidophorus Duméril & Bibron, 1839
* *Tropidophorus beccarii* Peters, 1871
* *Tropidophorus brookei* (Gray, 1845)
* *Tropidophorus iniquus* van Lidth de Jeude, 1905
* *Tropidophorus micropus* van Lidth de Jeude, 1905
* *Tropidophorus mocquardii* Boulenger, 1894
* *Tropidophorus perplexus* Barbour, 1921

VARANIDAE
Varanus Merrem, 1820
Varanus dumerilii (Schlegel, 1839)
Varanus rudicollis Gray, 1845
Varanus salvator (Laurenti, 1768)

Aphaniotis fusca (Peters, 1864)

SVL to 67 mm.

A slender lizard from lowland forests, with a reduced nuchal crest; long and slender limbs; dorsum dark brown; venter paler; two dark interorbital bars; and inner lining of mouth dark blue. Diet comprises caterpillars, beetles, millipedes, cockroaches and termites and 1–2 eggs are produced.

Distribution: Southern Thailand, the Malay Peninsula, Sumatra, Simalur, Nias (Mentawai Islands), Borneo, Singkep and the Natuna Islands.

Bronchocela cristatella (Kuhl, 1820)

SVL to 130 mm.

A familiar tree lizard from parks and gardens, as well as lowland forests and the midhills; body compressed; nuchal crest with elongate scales; dorsal crest somewhat distinct, dorsum green, sometimes with white or light blue spots or bars, changeable to brown. Diet comprises beetles, flies and ants and 1–4 eggs are produced at a time, each spindle-shaped, with pointed ends.

Distribution: Southern Myanmar, Thailand, the Nicobar Archipelago, the Malay Peninsula, Sumatra, Borneo, Java, the Lesser Sundas, Makulu and the Philippines.

Bronchocela jubata
Duméril & Bibron, 1837

SVL to 150 mm.

A rather rare (in Borneo) tree lizard from Kalimantan; body relatively robust; nuchal and dorsal crest with elongated scales; dorsum green, changeable to brown or black, with yellow or red spots or vertical bars. Inhabits lowland forests as well as disturbed areas; diet comprises insects and two eggs are laid at a time, although more than a clutch may be produced in a year.

Distribution: Java, Borneo, Bali, Singkep, Sulawesi, Karakelang and Salibabu Archipelagos.

Draco cornutus Günther, 1864

SVL to 85 mm.

A beautiful flying lizard; body slender, with a thorn-like scale over eye; tympanum scaleless; dewlap triangular, covered with small scales; nostril oriented laterally; dorsal crest absent; dorsum bright green to greenish-brown, in males; tan or light brown in females; patagium reddish-orange with dark spots or bands; a dark interorbital spot. Inhabits the plains and midhills, and feeds exclusively on small black ants. Clutches comprise 3–4 eggs.

Distribution: Sumatra, Borneo, Java, Bangunan Islands and the Sulu Archipelago.

11

Draco fimbriatus **Kuhl, 1820**

SVL to 132 mm.

A large flying lizard; body relatively robust; spinous projection over eye; tympanum large, scaleless; males with a low nuchal sail; dorsum and patagium greyish-brown, with grey and pale green markings. Inhabits lowlands and mid-elevation forests. Diet unknown and presumably comprises arthropods and 2–4 eggs are laid at a time.

Distribution: Southern Thailand, the Malay Peninsula, including Singapore, Sumatra, the Mentawai Archipelago, Borneo, Java and Mindanao in the Philippines.

***Draco haematopogon* Boie in Gray, 1831**
SVL to 94 mm.

A slender flying lizard; tail crest absent; tympanum large, skin-covered; dewlap covered with small scales; no thorn-like scale above eyes; dorsum olive or brownish-grey, with indistinct lighter and darker spots; patagium black with yellow spots. Inhabits midhills and submontane forests, and diet probably comprises ants and other small insects. Between 2–3 eggs are produced at a time.

Distribution: The Malay Peninsula, Sumatra, Java and Borneo.

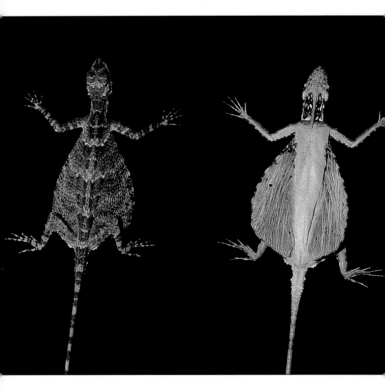

Draco maximus **Boulenger, 1893**

SVL to 139 mm.

A large, robust flying lizard; no spinous projection above eye; males with nuchal sail; dewlap covered with small scales, dorsum green, with a brownish-olive pattern of bands; patagium black with discontinuous olive-brown lines. Inhabits river-edges from the lowlands to about 1000 m and diet presumably comprises ants and other insects. Between 1–5 eggs are produced at a time.

Distribution: The Malay Peninsula, Sumatra, Borneo and the Natuna Islands.

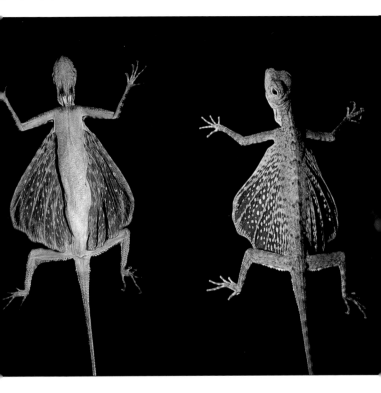

Draco melanopogon **Boulenger, 1887**

SVL to 93 mm.

A slender flying lizard; spinous projects above eye absent; dewlap elongate, scales covering dewlap slightly enlarged; dorsum olive or green with brownish-grey bands or diamond-shaped spots; patagium black with scattered yellow-orange spots, Inhabits lowland forests. Diet comprises ants, beetles, millipedes, isopods and termites and two eggs are produced.

Distribution: Peninsular Thailand, the Malay Peninsula, including Singapore, Sumatra, Borneo and the Natuna Islands.

Draco quinquefasciatus Hardwicke & Gray, 1827

SVL to 110 mm.

A slender flying lizard; no spinous projections above eye; dewlap tapering to a narrow tip; males with a low nuchal sail; dorsum bright green in males, brownish-olive in females, with dark specklings; patagium yellow or orangish-red above, with five dark brown or black cross-bars. Inhabits forests from the lowlands to the midhills. Feeds exclusively on ants and 1–4 eggs are produced.

Distribution: Southern Thailand, the Malay Peninsula, including Singapore, Sumatra, Pulau Sinkep, Pulau Belitung and Borneo.

Draco sumatranus
Schlegel, 1844

SVL to 85 mm.

The commonest Bornean flying lizard, a slender species; tail crest absent; dewlap triangular, covered with small scales; nuchal crest present, males with blue forehead when displaying; dorsum light brown, with dark brown blotches; dewlap bright yellow, with black dots at base. Inhabits open forests, plantations, parks and gardens. Diet includes ants and termites and 1–5 eggs are produced at a time.

Distribution: Thailand, the Malay Peninsula, including Singapore, Sumatra, the Mentawai and Riau Archipelagos, Borneo and Palawan.

Gonocephalus bornensis (Schlegel, 1848)

SVL to 136 mm.

A robust arboreal lizard; nuchal and dorsal crests continuous, highly developed in males; dorsum bright green with five dark bands; sides of head and flanks green spotted; sides of body with light oval spots; nuchal and body crest brown and yellow; dewlap pale with dark, broken stripes. Inhabits primary rainforests in the midhills. Diet comprises ants and spiders and a clutch size of four eggs is known.

Distribution: Endemic to Borneo.

Gonocephalus doriae (Peters, 1871)

SVL to 163 mm.

A brightly coloured forest lizard; nuchal crest composed of low, overlapping and slightly crescentic scales; ridge over head distinctly raised; no spine-like scales on crest in adult males; dorsum green, changeable to reddish-brown with dark and light flecks; gular sac grey with dark stripes. Inhabits lowland rainforests, a rather rare (in Borneo) tree lizard from Kalimantan; body relatively robust; nuchal and dorsal crest with elongated scales; dorsum green, changeable to brown or black, with yellow or red spots or vertical bars. Inhabits lowland forests as well as disturbed areas; diet comprises insects and two eggs are laid at a time, although more than a clutch may be produced in a year.

Distribution: Java, Borneo, Bali, Singkep, Sulawesi, Karakelang and Salibabu Archipelagos.

Gonocephalus liogaster (Günther, 1872)

SVL to 140 mm.

Another large forest-dwelling lizard; nuchal and dorsal crests continuous; males brown or green, with dark reticulate pattern on flanks; females with yellow cross-bars; eyes of males bright blue, the skin surrounding the orbit reddish-orange; in females, eyes are brown. Inhabits lowland rainforests, and also peat swamp forests. Diet comprises insects and clutches of 1–4 eggs are produced at a time.

Distribution: Peninsular Malaysia, Sumatra and Borneo.

Gonocephalus mjobergi **Smith, 1925**

SVL 88 mm.

A poorly-known tree lizard from Gunung Murud, Sarawak; body robust; supraciliary border not raised; dewlap small, its edge feebly serrated and covered with small scales; nuchal crest present; dorsal crest a small ridge; dorsum pale green, changeable to brownish-grey, with narrow grey reticulate pattern, which, on lower flanks, encloses yellow spots. Known only from montane forests. Diet presumably comprises insects. Breeding habits unknown.

Distribution: Endemic to Borneo.

Phoxophrys borneensis Inger, 1960

SVL to 155 mm.

A small shrub-dwelling lizard; body short, squat; spine above eyes absent; dorsum brown to greyish-brown, with yellowish-tan bands; two dark interorbital bars, the anterior narrower; upper lip cream. Inhabits montane forests between 1300–1800 m. Diet presumably comprises insects and two eggs are produced at a time.

Distribution: Endemic to Borneo (Sarawak, Sabah and Kalimantan).

Phoxophrys cephalum (Mocquard, 1890)

SVL to 84 mm.

Another lizard from high elevation forests; body short, squat; supraciliary spines absent; nuchal crest comprises 7–8 thick, conical scales; dorsum pale green, with dark green or greyish-green wavy bands, changeable to dark brown. It inhabits submontane and montane forests, at elevations between 1300–2100 m. Of its diet or reproductive habits, nothing is on record.

Distribution: Endemic to Borneo (Sabah).

Phoxophrys nigrilabris (Peters, 1864)

SVL to 58 mm.

A short, squat lizard from shrubs in forests in the lowlands; spine above eyes absent; nuchal crest comprises compressed scales; gular scales distinctly keeled; dorsum of adult males brown with transverse blue bands; females and juveniles brown to olive; ventrally cream. Inhabits lowland dipterocarp forests, its diet includes insects and other arthropods. Reproductive habits unknown.

Distribution: Borneo, and apparently also, Pulau Sirhassen, in the Natuna Archipelago.

Phoxophrys spiniceps Smith, 1925

SVL to 60.3 mm.

A montane lizard from shrubs, known only from Gunung Murud in Sarawak; body short, squat; enlarged spine above eyes; gular scales keeled; dorsum greenish-grey with brown patches, change-able to brown, with thin, pale transverse lines meeting at vertebral region. Inhabits high altitudes of north-central Sarawak, and adjacent regions of southern Sabah, in northern Borneo. Diet presumably comprises arthropods and two eggs are produced at a time.

Distribution: Endemic to Borneo.

Aeluroscalabotes felinus (Günther, 1864)

SVL to 122 mm.

A brightly-coloured forest-dwelling lizard; body slender; eyelids fleshy; tail rounded and capable of being curled laterally; dorsum reddish-brown, with white spots and vertebral stripe on body and tail. Inhabits lowland rainforests and peat swamp forests up to about 800 m; diet comprises arthropods and one to two elongate eggs with parchment shells are produced.

Distribution: Peninsular Thailand, the Malay Peninsula, Sumatra, Borneo.

Cnemaspis kendallii (Gray, 1845)

SVL to 80 mm.

A forest-dwelling, day-active gecko; body slender; canthal ridge developed; tail with a median row of pointed scales below; dorsum pale brown, with dark brown oblong spots forming  seven interrupted bands. Inhabits lowland forests and peat swamps, its diet includes ants, earthworms, beetles and millipedes. Two eggs are produced at a time.

Distribution: The Malay Peninsula, including Pulau Tioman, the Riau and Natuna Archipelagos and north-western Borneo.

27

Cnemaspis nigridia **(Smith, 1925)**

SVL to 69.8 mm.

A rock-dwelling gecko, from Gunung Gading and the adjacent hills of western Sarawak; body robust; canthal ridge developed; tail without a median row of pointed scales below; dorsum brownish-olive, with black blotches. Inhabits granite and limestone hills of western Sarawak. Diet comprises spiders and presumably other arthropods. Two eggs are produced at a time.

Distribution: Endemic to north-western Borneo and Natuna Island.

Cosymbotus craspedotus (**Mocquard, 1890**)

SVL to 62 mm.

A gliding gecko from lowland rainforests; body slender; body and tail depressed; dorsum with scattered tubercles; skin frills on sides of body, tail, sides of throat and along lateral edges of limbs; digits nearly fully webbed; dorsal surface greyish-brown, with two rows of dark, rectangular spots. Inhabits trees in rainforests and is insectivorous. Its reproductive habits are unknown.

Distribution: Peninsular Thailand, the Malay Peninsula and Borneo, and probably also, Java.

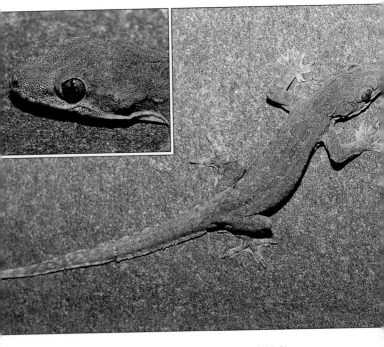

Cosymbotus platyurus (Schneider, 1792)

SVL to 69 mm.

A familiar house gecko; body slender; a fringe of skin on sides of body and back of hind limbs; tail flattened, with a serrated margin; body depressed, smooth, with tiny granules; dorsum light grey, with a dark grey streak between eye and shoulder. A human commensal, abundant in towns and cities, but also known from forests. Diet comprises spiders and ants. Two eggs are produced at a time, several clutches being laid every year.

Distribution: Eastern India, Nepal, Andaman and Nicobar Islands, Sri Lanka, southern China, and all of south-east Asia to Sulawesi and the Philippines; introduced into Papua New Guinea.

Cyrtodactylus baluensis (Mocquard, 1890)
SVL to 86 mm.

A tree gecko from northern Borneo; body slender; preanal groove absent; dorsum brown to yellowish-brown with irregular dark spots that may form dark cross-bars; head with dark brown lateral band from snout-tip to nape; limbs and tail dark-banded. Inhabits dipterocarp rainforests to montane oak forests, between 150–2500 m. Diet comprises insects and other arthropods and two eggs are produced at a time.

Distribution: Endemic to northern Borneo.

Cyrtodactylus cavernicolus Inger & King, 1961

SVL to 80.8 mm.

A gecko from limestone regions of Niah and Mulu, in Sarawak; body slender; preanal grooves containing two pairs of pores; dorsal surface covered with small, granular scales; dorsum brown, with dark-edged brown cross-bars, changeable to brown-black; a dark stripe from corner of eyes to nape. Inhabits caves and forests of northern Sarawak; diet comprises flattened cave cockroaches and moths that dwell in guano. Reproductive habits unknown.

Distribution: Endemic to northern Borneo.

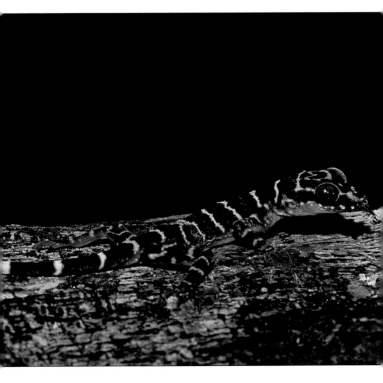

Cyrtodactylus consobrinus **(Peters, 1871)**

SVL to 125 mm.

A large, forest gecko; body robust; dorsum with scattered tuberculate scales; forehead with pale narrow network of reticulations; dorsum dark chocolate brown with 4–8 white or yellow transverse bands. Inhabits lowland dipterocarp forests up to 1100 m, and also caves; diet comprises insects. Two eggs are produced several times a year.

Distribution: The Malay Peninsula, Sumatra, Sinkep and Borneo.

Cyrtodactylus ingeri Hikida, 1990

SVL to 80.2 mm.

A forest gecko from northern and north-western Borneo; body slender; dorsum with large, tuberculate scales; dorsum grey or light yellow-brown, with 5–6 dark brown bands, that may be broken up to form paravertebral blotches; a dark stripe from posterior edge of eye to insertion of forearm. Inhabits riparian forests, between 500–800 m, its diet presumably comprises arthropods and two eggs are produced at a time.

Distribution: Endemic to Borneo (Sabah and Brunei Darussalam).

Cyrtodactylus matsuii Hikida, 1990

SVL to 105 mm.

A montane gecko from Sabah; body stout; dorsum yellowish-brown or pale brown, with irregular dark cross-bars; forehead with

small dark spots; dark band on interorbital region joining posterior edge of eyes. Inhabits forests at elevations between 900–1600 m, its diet comprises insects and other arthropods. Nothing is on record of its reproductive habits.

Distribution: Endemic to northern Borneo (Sabah).

Cyrtodactylus pubisulcus Inger, 1957

SVL to 77 mm.

A common lowland gecko from Borneo; body slender; preanal groove present; dorsum grey, dark markings on dorsal surface arranged in the form of cross-bars or blotches, sometimes arranged in a longitudinal series. Inhabits lowland rainforests and peat swamps; its diet comprises insects such as cockroaches and two eggs are produced at a time.

Distribution: Endemic to Borneo (Sarawak and Brunei Darussalam).

Cyrtodactylus quadrivirgatus **Taylor, 1962**

SVL to 71 mm.

A lowland gecko from north-western Borneo; body slender; dorsal tubercles arranged in regular longitudinal rows; dorsum grey or dark brown, with four black longitudinal lines, separated by lighter areas; no continuous black band joining the eyes. Inhabits primary and secondary forests, its diet comprises arthropods. Two eggs are produced at a time.

Distribution: Southern Thailand, the Malay Peninsula, northern Sumatra, the Mentawai Archipelago and north-western Borneo.

Cyrtodactylus yoshii
Hikida, 1990

SVL to 96 mm.

A lowland gecko from Sabah; body robust, subdigital lamellae not enlarged; dorsum grey, with five dark V-shaped cross-bars on body; a dark brown V-shaped stripe on nape; and venter unpatterned, pale brown; its diet presumably comprises large insects and reproductive habits unknown.

Distribution: Endemic to northern Borneo.

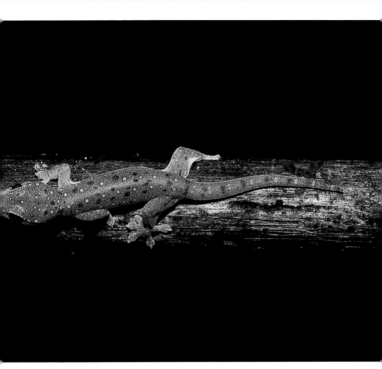

Gehyra mutilata (Wiegmann, 1834)

SVL to 64 mm.

A common house gecko, with a delicate skin; body slender; head relatively large and oval; dorsum pale grey, usually with a paler vertebral area, with dark and light spots. Inhabits both human habitations and primary forests, its diet comprises insects and isopods and one to two eggs are produced at a time.

Distribution: Mainland India, Andaman and Nicobar, Sri Lanka; to south-east Asia, east to New Guinea and the Philippines. Introduced into Mauritius, Seychelles, Madagascar, México, Cuba and Hawai'i.

Gekko gecko
Linnaeus, 1758
SVL to 176 mm.

A noisy tree gecko from islands to the north of Borneo; body robust; head relatively large; dorsum slaty grey to bluish-grey, with red or orange spots. Mostly restricted to offshore islands, such as those north of Sabah; diet includes moths, grasshoppers, beetles, spiders, other geckos, small mice and snakes; and 1–2 eggs are laid in tree holes.

Distribution: Eastern India, Nepal, Bangladesh, east to southern China and south-east Asia, to the Philippines; introduced into Florida and Hawai'i, Martinique in the West Indies, and Madagascar.

Gekko monarchus (Duméril & Bibron, 1836)

SVL to 102 mm.

A large, rough house gecko; body robust and tuberculate; dorsum greyish-brown, with dark brown blotches arranged in 7–9 pairs; inhabits both buildings and forest edges up to 1500 m and diet comprises insects and other invertebrates. Two are produced at a time.

Distribution: Southern Thailand, the Malay Peninsula, Sumatra, the Mentawai Archipelago, Pulau Simeulue, Borneo, Java, Maluku and the Philippines.

Gekko smithii (Gray, 1842)

SVL to 180 mm.

A large forest gecko, whose barking call is distinctive; body robust; dorsum greyish-brown, with a transverse series of white spots, inhabits forested habitats in the lowlands as well as the midhills, its diet comprises insects, such as grasshoppers and two eggs are laid at a time.

Distribution: The Nicobar Islands, southern Thailand, the Malay Peninsula, Sumatra, Borneo, Pulau Nias and Java.

Hemidactylus brookii Gray, 1845

SVL to 63 mm.

A gecko from disturbed habitats, and the Bornean population probably the result of unsuccessful human introduction; body robust, flattened; with rows of tubercles; tail with spiny tubercles; dorsum dark brown to light grey; dark spotted; two dark lines along nostrils and eyes. Inhabits parks, gardens, houses, as well as forests; diet comprises small insects and two eggs are produced at a time.

Distribution: Tropical Africa to southern China and south-east Asia.

Hemidactylus frenatus **Duméril & Bibron, 1836**
SVL to 67 mm.

A common house gecko from towns and cities in Borneo. Body slender, depressed; dorsal scales smooth; dorsum yellowish-brown to almost black, with darker markings; a light-edged brown streak along sides of head. Inhabits man-made structures as well as forested areas, its diet includes insects and spiders and two eggs are produced at a time.

Distribution: Widespread from India, Sri Lanka, southern China, to south-east Asia and introduced into Central and South America, Madagascar, eastern and southern Africa, Mauritius, New Guinea, Polynesia and Australia.

Hemidactylus garnotii **Duméril & Bibron, 1836**

SVL to 65 mm.

An all-female parthenogenetic species, rare in Borneo; body slender and depressed; tail slender, depressed, with denticulate lateral edges dorsum brownish-grey, sometimes marbled with brown. Inhabits primary forests and buildings, its diet presumably consists of on insects and other arthropods, and two eggs are produced at a time.

Distribution: Eastern India, southern China, east through south-east Asia to the South Pacific, Solomons, Fiji and Tahiti; introduced into Florida and Hawai'i.

***Hemiphyllodactylus typus* Bleeker, 1860**

SVL to 47 mm.

A worm-like, tree-dwelling gecko; body slender and depressed; limbs reduced; tail slender, prehensile, dorsum brown; a dark brown stripe from nostril to shoulder; a light V-shaped mark at base of tail. Inhabits lowland forests and mangroves; diet comprises small insects. An all-female species, producing two eggs without mating.

Distribution: Mauritius, Nicobar Islands, Sri Lanka, Myanmar, China, east through south-east Asia to New Guinea, Solomon Islands, New Caledonia, Polynesia, Hawai'i and the Mascarene Islands.

Lepidodactylus lugubris (Duméril & Bibron, 1836)
SVL to 49 mm.

Another all-female species of tree lizard, rare in Borneo; body slender, with a wide tail; dorsum cinnamon or greyish-brown, with darker flecks; a dark stripe along face; cross-bars on body. Inhabits lowland forests and mangroves, its diet includes insects, although nectar and plant juice are also lapped up; one or two eggs are produced without mating.

Distribution: Maldives, Andaman and Nicobar Islands, Sri Lanka, southern China, east through south-east Asia to New Guinea, and the South Pacific; introduced into Panama, Ecuador, Galapagos Islands and Central America.

Lepidodactylus ranauensis **Ota & Hikida, 1988**

SVL to 47.7 mm.

A house gecko from the hills of northern Borneo; body slender, with a wide tail; dorsum greyish-brown; a pair of dark triangular markings on side of tail base. Only known from walls of buildings, its diet presumably comprises insects and two eggs are produced at a time.

Distribution: Endemic to Borneo (Sabah).

Luperosaurus browni **Russell, 1979**

SVL to 66.5 mm.

A tree-dwelling gecko; body slender; digits half-webbed; skin folds present on limbs; tail depressed with lateral spines; dorsum light grey, with minute black spots on head, body and limbs. Inhabits lowland rainforests and diet presumably comprises small insects. Two eggs are glued to leaves.

Distribution: The Malay Peninsula and northern and central Borneo.

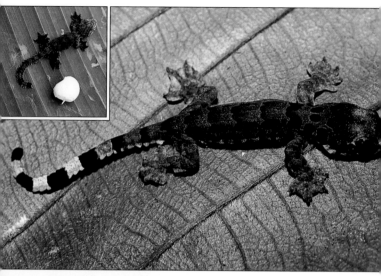

Ptychozoon horsfieldii
(Gray, 1827)

SVL to 80 mm.

A parachuting gecko from the lowlands; body robust; dorsal tubercles absent; tail lobe size reduction to tail tip gradual; 21–22 denticulate tail lobes; dark circular nuchal spots and four transverse dark bands on trunk present. Inhabits lowland rainforests as well as man-made structures, its diet presumably comprises small insects and two eggs are attached to branches.

Distribution: Myanmar, Thailand, the Malay Peninsula, Sumatra and Borneo.

50

Ptychozoon kuhli
Stejneger, 1902

SVL to 107.8 mm.

Another parachuting gecko from the lowlands; body robust; tail tip ending in a broad flap; dorsum grey or reddish-brown with 4–5 wavy dark brown transverse bands. Inhabits large trees in lowland forests and occasionally, enters walls of houses, its diet includes arthropods, including grasshoppers and two eggs are glued to tree trunks and branches.

Distribution: Southern Thailand, the Malay Peninsula, Sumatra, the Mentawai Archipelago, Java and Borneo.

51

Takydromus sexlineatus Daudin, 1802

SVL to 61 mm.

A long-tailed lizard from grasslands; body slender; head long; dorsal surface with large, smooth, plate-like scales; tail between 3–5 times as long as body; dorsum brown or olive-brown, with a green stripe from eyes to base of tail. Inhabits grasslands and marshes, it feeds on insects and millipedes and clutches of 2–3, eggs are produced at a time.

Distribution: North-eastern India, Myanmar, Thailand, Vietnam, southern and eastern China, Borneo, Sumatra, Java.

Lanthanotus borneensis **Steindachner, 1877**

SVL to 200 mm.

An enigmatic lizard, endemic to Borneo, whose closest relative are the monitor lizards; body slender, elongate and cylindrical, with short limbs; head blunt; dorsum unpatterned brownish-orange, or with a dark vertebral stripe. Inhabits lowlands near streams, and also agricultural lands, diet comprises earthworms and crustaceans and 2–5 eggs are produced at a time.

Distribution: Known only from isolated localities in Sarawak and Kalimantan.

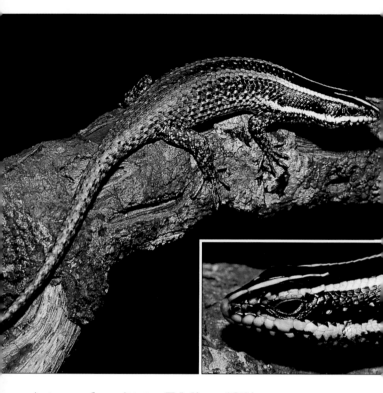

Apterygodon vittatus **Edeling, 1864**

SVL to 96 mm.

A familiar tree skink in the lowlands; body robust; dorsals keeled; head and anterior of body black; rest of dorsum brownish-grey; with dark and light spots; a light cream or yellow stripe from snout-tip to back of head; a pale stripe from above eyes to along body. Inhabits lowland rainforests as well as parks and gardens; diet is primarily ants, also eaten are other small insects; and 2–4 eggs are produced at a time.

Distribution: Endemic to Borneo.

Dasia grisea (Gray, 1845)

SVL to 130 mm.

A dark coloured tree skink from closed canopy forests; body slender; three strong keels on dorsal scales; dorsum light or dark brown with 8–14 narrow dark rings. Inhabits lowland dipterocarp forests and feeds on ants, termites, beetles, snails, as well as fruits; clutches of 2–6 eggs are produced.

Distribution: Malay Peninsula, Borneo, Sumatra and the Philippines.

***Dasia olivacea*
Gray, 1839**

SVL to 115 mm.

Another tree skink, this one inhabits open forests; body robust; 3–5 weak keels on dorsal scales; dorsum olive to greenish-brown. Inhabits forests in the lowlands, up to 1,200 m; diet comprises bees, beetles, ants and flies, and other arthropods; and 6–14 eggs are laid.

Distribution: Myanmar, Thailand, Cambodia, Peninsular Malaysia, Singapore, Sumatra, Borneo, Java, Natuna and Nicobar Islands.

Emoia atrocostata (Lesson, 1830)

SVL to 97.5 mm.

A common beach- and mangrove forest lizard; body slender; limbs and tail well developed; dorsum greyish-olive, flecked with dark brownish-grey; inhabits coastal regions, including both sandy and rocky beaches; diet comprises small crabs, termites, fishes and other lizards; and 1–3 eggs are produced at a time.

Distribution: Malay Peninsula, Sumatra, Borneo, Java, east to the Philippines, New Guinea, the Solomons and northern Australia.

C.L. Chan

Emoia caeruleocauda (De Vis, 1892)

SVL to 50.9 mm.

A sea-shore dwelling lizard with a blue tail; body robust; snout short, tapering; dorsal surface of body with a dark vertebral stripe in males, yellow in females, and undersurface of tail blue; diurnal and terrestrial; two eggs are produced at a time and hatchlings measure 22.4 mm in SVL.

Distribution: Islands to the east of Sabah; also, Sulawesi, New Guinea, east to Fiji and the Solomon Islands.

Lamprolepis vyneri (**Shelford, 1905**)

SVL to 66 mm.

A tree-dwelling lizard known only from two localities from the low hills of central Sarawak; body slender; snout obtusely pointed; body scales smooth; dorsum olive-grey, some scales edged with black, forming four longitudinal stripes, flanks with black-edged brown scales; inhabits the midhills; nothing is known of its diet or reproductive habits.

Distribution: Endemic to Borneo.

Lipinia **undescribed species**

SVL to 40.6 mm.

An undescribed species of skink, previously identified as a species from the Philippines; body slender; snout acute; external ear opening absent; dorsum tan brown with a series of dark grey-brown stripes; paired paravertebral stripes from behind eye to tail tip; paired dorsal stripes from temporals to inguinal region. Known from isolated lowland localities in Borneo, its diet remain unknown, and two eggs are produced at a time.

Distribution: Endemic to Borneo.

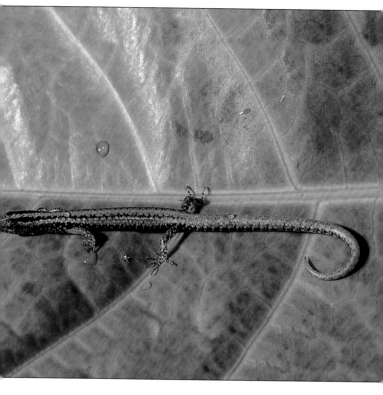

***Lipinia nitens* (Peters, 1871)**

SVL to 33.6 mm.

A poorly-known, terrestrial skink restricted to north-western Borneo; body slender; snout pointed; dorsal and lateral scales smooth; external ear opening absent; dorsum yellowish-green; sides spotted with black and green; pale yellow vertebral stripe, with jagged edged black lines. Inhabits lowland forests and diet includes ants. Its reproductive habits remain unknown.

Distribution: Isolated localities in western Sarawak.

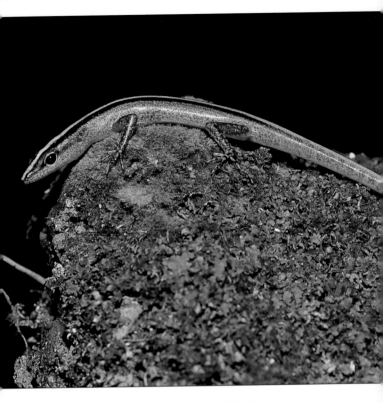

Lipinia vittigera (**Boulenger, 1894**)

SVL to 42 mm.

   A common but rarely seen tree-dwelling skink; body slender; snout elongated; ear opening small; dorsum brownish-black, with a bright yellow vertebral stripe, commencing from snout-tip. Inhabits lowland forests; its diet comprises small insects and clutches of 2–4 eggs are laid at a time.

   Distribution: Southern Myanmar, Thailand, Laos, Malay Peninsula, Mentawai Archipelago, Sumatra and Borneo.

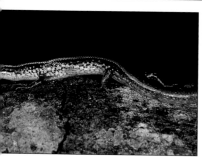

Lygosoma bowringii (Günther, 1864)

SVL to 58 mm.

A common, garden skink, frequently seen in urban areas; body slender, elongate; head scarcely distinct from neck; limbs reduced; dorsum bronze brown; with dark sides. More abundant in disturbed areas than in forests; its diet comprises small insects and clutches of 2–4 eggs are produced at a time.

Distribution: Myanmar, Andaman Islands, Thailand, the Malay Peninsula, Borneo, Java, Sulawesi, possibly Sumatra, Vietnam, Cambodia, Laos and the Sulu Archipelago of the southern Philippines; introduced into the Christmas Island, Indian Ocean.

Mabuya indeprensa Brown & Alcala, 1980

SVL to 67 mm.

A terrestrial skink from open forests and other disturbed areas; body robust; dorsal scales with three keels; dorsum brownish-tan to light brown, typically with 2–4 longitudinal rows of small dark brown blotches; throat greyish-blue. Inhabits lowland forests, and often found in the vicinity of human settlements; diet presumably comprises insects; reproductive habits unknown.

Distribution: The Philippines and Sabah in north-eastern Borneo.

Mabuya multifasciata **(Kuhl, 1820)**

SVL to 137 mm.

A familiar terrestrial skink, widespread in Borneo; body robust; dorsal scales with three, rarely five, keels; dorsum bronze brown, with an orange or reddish-orange lateral band. Inhabits disturbed habitats, including forest clearings, its diet is primarily insectivorous, including cockroaches, isopods, spiders; smaller lizards; ovoviviparous, giving birth to 1–10 live young.

Distribution: Throughout Borneo, the distribution extends from southern China and north-eastern India, through Indo-China and the Malay Peninsula to the Greater and Lesser Sundas and Philippines. Records from New Guinea may be through human introductions.

Mabuya rudis Boulenger, 1887

SVL to 120 mm.

A small terrestrial forest skink, with a rough forehead; body robust; forehead scales at posterior rugose; dorsal scales with three strong keels; dorsum olive-brown, with a light-edged dark brown line on the side; throat of adult males crimson. Inhabits forests in the lowlands and midhills up to elevation of about 1,300 m; its diet includes grasshoppers, cockroaches, moths, flies and isopods; and clutches of 2–4 eggs laid.

Distribution: Sumatra, Mentawai Archipelago, Borneo, Nicobar Islands, the Sula Archipelago, Sulawesi and Sulu Archipelago.

Mabuya rugifera (Stoliczka, 1870)

SVL to 65 mm.

An semi-arboreal skink from the lowlands; body robust; dorsal scales with five, rarely seven, distinct keels; dorsum blackish-brown, with 5–7 greenish-cream longitudinal stripes, sometimes broken up to form spots. Inhabits forests in the midhills; diet comprises insects. Reproductive habits unknown.

Distribution: Southern Thailand, Nicobar Islands, Malay Peninsula, Sumatra, Borneo, Java and Bali.

Sphenomorphus aesculeticola Inger, Tan, Lakim & Yambun, 2002

SVL to 42 mm.

A small terrestrial/fossorial skink with reduced limbs; body slender; dorsum brown, with many scales dark spotted, forming a series of dark lines or checkered pattern; a dark lateral band. Inhabits montane regions of northern Borneo. Diet presumably comprises small insects and their larvae, and two eggs are produced at a time.

Distribution: Restricted to Mount Kinabalu and adjacent mountains.

Sphenomorphus cyanolaemus **Inger & Hosmer, 1965**
SVL to 60 mm.

A blue-throated ground skink that can also climb trees; body slender; limbs relatively long; dorsum bronze- or olive-brown, with two rows of dark spots; dark dorsolateral stripe. Inhabits lowland forests, its diet presumably comprises insects; two eggs are produced at a time, and may be deposited in ant heaps.

Distribution: Malay Peninsula, Sumatra and Borneo.

Sphenomorphus haasi Inger & Hosmer, 1965

SVL to 57 mm.

A small; terrestrial skink from western Sarawak; body slender; ear opening lacking lobules; dorsum greyish-brown, with pale olive blotches, pale blue sclera of eye; dark dorsolateral band absent. Inhabits lowland forests of north-western Borneo, its diet presumably comprises small arthropods. Reproductive habits unknown.

Distribution: Endemic to Borneo.

Sphenomorphus kinabaluensis **(Bartlett, 1895)**
SVL to 58 mm.

A montane skink found commonly in Gunung Kinabalu; body slender; dorsum light to dark brown, with several longitudinal rows of dark brown to yellow spots and occasionally also, dark brown speckles; a black dorsolateral stripe with small yellowish flecks. Inhabits altitudes between 1,600–2,200 m; its diet comprises insects and 1–2 eggs are produced, sometimes within ant nests.

Distribution: Restricted to northern Borneo.

Sphenomorphus multisquamatus **Inger, 1958**
SVL to 68.5 mm.

A large, terrestrial skink, widespread in western Borneo; body robust; dorsum dark greyish-brown, with 2–4 rows of squarish black spots, with or without dark dorsolateral bands. Inhabits lowland rainforests, its diet comprises small insects. Nothing is known of its reproductive biology.

Distribution: Endemic to Borneo.

Sphenomorphus murudensis **Smith, 1925**

SVL to 49.3 mm.

A montane skink from the Gunung Murud massif in Sarawak; body slender; snout rounded; lower eyelid scaly; dorsum dark brown, with black spots; a dark band on the sides. Inhabits summits at altitudes of over 1500–2400 m. Diet remain unknown, and two eggs are produced at a time.

Distribution: Endemic to Gunung Murud, in north-western Borneo.

Tropidophorus beccarii **Peters, 1871**

SVL to 98 mm.

A smooth-scales water skink from the low hills of Borneo; body robust in adults, slender in juveniles; scales smooth at least in adults; dorsum dark brown or reddish-brown, with dark brown blotches and cross-bars. Inhabits rocky streams within dipterocarp forests and diet comprises water insects. Ovoviviparous, giving birth to four live young.

Distribution: Endemic to Borneo.

Tropidophorus brookei (Gray, 1845)

SVL to 101 mm.

The commonest Bornean water skink; body robust in adults, slender in juveniles; dorsum dark brown with darker spots and blotches that may form transverse bands; a black spot on sides of neck; flanks with dark and white spots. Inhabits rocky streams in lowland dipterocarp forests, diet presumably aquatic arthropods and 1–5 live young ones are produced at a time.

Distribution: Endemic to Borneo.

***Tropidophorus micropus* van Lidth de Jeude, 1905**
SVL to 40 mm.

A poorly-known water skink from central and northern Borneo; body slender; scales on forehead distinctly striated; dorsum dark brown, with a black spot on sides of neck; and venter cream, with irregular dark spots. Inhabits rocky streams within lowland forests. Diet remains unknown, and likely to be aquatic arthropods and three live young ones are produced at a time.

Distribution: Endemic to Borneo.

Tropidophorus mocquardii Boulenger, 1894
SVL to 95 mm.

A high-elevation water skink, from northern Borneo; body slender; forehead scales smooth; dorsum brown, with dark transverse bands; flanks with white spots; and venter cream. Inhabits the midhills of Mount Kinabalu. Its diet and reproductive habits remain unstudied.

Distribution: Endemic to northern Borneo.

Tropidophorus perplexus **Barbour, 1921**

SVL to 73 mm.

A poorly-known water skink from the interior of Sarawak and southern Sabah; body slender; scales on forehead rugose; dorsum rich brown, with paler narrow cross-bars; and venter yellow. Inhabits slow-flowing streams of northern and central-northern Borneo. Diet unknown and presumably aquatic arthropods. Reproductive habits unknown.

Distribution: Endemic to north-central Borneo.

Varanus rudicollis Gray, 1845

SVL to 1.46 m.

A rough-necked tree monitor; body slender; snout relatively long; nuchal scales strongly keeled; dorsum almost black in adults, with yellow tinge on neck and foreparts of body in juveniles. Inhabits lowland forests, its diet comprises ants, termites, stick insects, cockroaches, grasshoppers, spiders, scorpions, and also, small mammals, frogs, fishes and crabs, and between 13–14 eggs are produced.

Distribution: Southern Myanmar, Thailand, the Malay Peninsula, Sumatra, Pulau Bangka, Borneo and probably the Philippines.

Varanus salvator (Laurenti, 1768)

SVL to 3 m.

A large monitor, commonly seen near waterbodies; body robust in adults, slender in juveniles; snout depressed; juveniles dark dorsally with, yellow spots. Inhabits many habitat conditions, both natural and man-made, and frequently seen in urban settings; diet comprises any animal matter, including carrion; between 5–30 eggs are produced.

Distribution: Sri Lanka and India, southern China, east through south-east Asia up to the Lesser Sundas and the Philippines.

Acknowledgements

Preparation of this photographic guide was supported by Universiti Malaysia Sarawak (UNIMAS) research grants (UNIMAS 120/98 [9], 192/99 [4], 1/26/303/2002 [40] and 01/59/376/2003 [113]), as well as an Intensification of Research in Priority Areas (IRPA) grant (08-02-09–1007-EA001), administered by the Institute of Biodiversity and Environmental Conservation (IBEC), UNIMAS. I thank Andrew Alek Tuen and Fatimah Abang, current and former Directors, IBEC, for support and facilities and current and past colleagues at IBEC/UNIMAS (S.J. Davies, D.S. Hill, and M.A. Rahman) and at Universiti Brunei Darussalam (J.K. Charles, D.S. Edwards, D.T. Jones, H. Pang, S. b. Nyawa and A.S. Kamariah).

Many additional colleagues supported my research on the herpetofauna of Borneo, among them: K. Adler, C.C. Austin, A. M. Bauer, R. Crombie, P. David, A. Greer, T. Hikida, M. Hoogmoed, R. Inger, D. Iskandar, K.K.P. Lim, C. McCarthy, U. Manthey, H. Ota, R. Stuebing, F.L Tan, H. Voris, V. Wallach and G. Zug.

As usual, I am indebted to my publisher and friend, Datuk Chan Chew Lun, for publishing this work and to Cheng Jen Wai for page layout.

Further Reading

Bauer, A.M. (1998). Lizards. In: H.G. Cogger & R. G. Zweifel (eds.). *Encyclopedia of reptiles and amphibians.* Pp: 126–173. Academic Press, San Diego.

Boulenger, G.A. (1885a). Catalogue of lizards in the British Museum (Natural History). Second edition. Vol. 1. Geckonidae (sic), Eublepharidae, Uroplatidae, Pygopodidae, Agamidae. British Museum (Natural History), London. xii + 436 pp + Pls. I–XXXII.

Boulenger, G.A. (1885b). Catalogue of lizards in the British Musum (Natural History). Second edition. Volume II. Iguanidae, Xenosauridae, Zonuridae, Anguidae, Anniellidae, Helodermatidae, Varanidae, Xantusiidae, Teiidae, Amphiesbaenidae. British Museum (Natural History), London. xiii + 497 pp + Pls. I–XXIV.

Boulenger, G.A. (1887). Catalogue of lizards in the British Museum (Natural History). Second edition. Volume III. Lacertidae, Gerrhosauridae, Scincidae, Anelytropidae, Dibamidae, Chamaeleontidae. British Museum (Natural History), London. xii + 575 pp + Pls. I–XL.

Brown, W.C. (1991). Lizards of the genus *Emoia* (Scincidae) with observations on their evolution and biogeography. *Mem. California Acad. Sci.* 15: i–vi + 1–94.

Das, I. (1996). Lizards. In: T. Whitten & J. Whitten (eds.). *Indonesian heritage. Wildlife. Volume 5.* Pp. 34–35. Editions Didier Millet/Archipelago Press, Singapore.

De Rooij, N. (1915). *The reptiles of the Indo-Australian Archipelago. Vol. I. Lacertilia, Chelonia, Emydosauria.* E.J. Brill, Leiden. xiv + 384 pp.

FURTHER READING

Diong, C.H. (1998). Lizards. In: H.-S. Yong (ed.). *The encyclopedia of Malaysia. Volume 3. Animals.* pp: 66–67. Archipelago Press, Singapore.

Hikida, T. (1980). (Lizards of Borneo.) *Acta Phytotaxonomica et Geobotanica* 31(1–3): 97–102. [In Japanese.]

Inger, R.F. (1983). Morphological and ecological variation in the flying lizards (genus *Draco*). *Fieldiana Zool. n.s.* 18: i–iv + 1–35.

Inger, R.F. and Tan Fui Lian (1996). *The natural history of amphibians and reptiles in Sabah.* Natural History Publications (Borneo), Kota Kinabalu. vi + 101 pp.

Malkmus, R., U. Manthey, G. Vogel, P. Hoffmann and J. Kosuch (2002). *Amphibians & reptiles of Mount Kinabalu (North Borneo).* Koeltz Scientific Books, Königstein. 424 pp.

Manthey, U. and W. Grossmann. (1997). *Amphibien and Reptilien Südostasiens.* Natur und Tier Verlag, Münster. 512 pp.

Tan, Fui Lian (1993). *Checklist of lizards of Sabah.* Sabah Parks Trustees, Kota Kinabalu. (2) + 18 pp.

LoveCats

LoveCats

Hugo Ross

BLACK & WHITE PUBLISHING

First published 2017
by Black & White Publishing Ltd
29 Ocean Drive, Edinburgh EH6 6JL

1 3 5 7 9 10 8 6 4 2   17 18 19 20

ISBN: 978 1 78530 085 1

Text © Black & White Publishing 2017

The publisher has made every reasonable effort to contact copyright holders of
images in this book. Any errors are inadvertent and anyone who for any reason has
not been contacted is invited to write to the publisher so that a full
acknowledgment can be made in subsequent editions of this work.

A CIP catalogue record for this book is available from the British Library.

Typeset by 3btype.com

Image used on p138 © iStock
All other images © Shutterstock

Introduction

Love is not necessarily the first thing that springs to mind when we think of cats. Even though we want to shower them with affection, we all know that cats can be temperamental, aloof and reserved. Some cats can be downright crazy! But while us humans often get given the cold shoulder, there is nothing cats love more than a cuddle with their furry friends.

Whether they are sunbathing together in the garden, snuggled on a blanket, climbing trees together, or keeping each other pristine, cats certainly have a lot of love for each other. If only they'd give us that much attention! *LoveCats* takes a look behind their cool exterior, capturing the cuddly kittens and affectionate felines that they really are.

This stunning compilation of photographs truly captures cats at their most adorable. Perfect for cat lovers, or simply lovers, this gorgeous collection is sure to melt any heart.

Ok, one more kiss!

This is our first
couple-shoot,
darling — put
some effort
into it!

We were lied
to - rolling
yarn back
into a ball is
way more
boring than
unravelling it!

Ok, now let's do a silly one!

Am I being given the silent treatment now?

And now we will reach around into the 'Sleeping Cat' pose. Inhale . . . and relax.

I just don't know why people think we're related?

Help me,
Scar!

13

I thought you
said I could
be little spoon
this time?

15

WAKE UP
AND GIVE ME
ATTENTION.

If this isn't
love, I don't
know what is.

Right,
bath time!

I'm really hoping you grow out of this habit of using me as a bed.

I sleep,
therefore
I cat.

Why did no one tell me there was a grey theme today?

Help –
trapped in
a snuggle
sandwich!

Muuuuum,
this is so
uncool.

Well, nature's nice and all, but we could be curled up on a beanbag in a warm living room.

Nope. No tuna = no love.

I have peeped outside our nest,
and that was enough activity
for one day.

We're starting a synchronised scowling team for the next cat olympics.

Welcome to
the cat pile!
The more
the merrier.

We are torn between wanting to love you and wanting to bite you.

41

We've fallen down
a soft, snuggly hole
and now we don't
want to get out.

I just really think if we put our heads together on this one we can come up with a way to get rid of the dog.

45

I'm sorry, can we help you with something?

Oooohhh shiny!

Spring has
spruuuung!

Pahahaha!!
Oh boy, your
one-liners
really crack
me up, Mr
Tiddlesworth!

53

Sometimes I really feel like I let you walk all over me.

Tonight is the night when two become one . . .

I feel like a lot of people are invading my personal space right now.

We're too cute to have ripped up the curtains . . .

61

Your scent is like a drug to me – like my own personal brand of fish-flavoured kitty treats.

We should not have
drunk so much last night.

I hope you have a good reason for waking us up?

What? No. We've never even heard of catnip.

Stoppp, it tickles!

Why you so far away?

Happiness
is a clean
forehead.

What do you mean it doesn't look like a heart?!

Who taught you how to kiss? Mike Tyson?

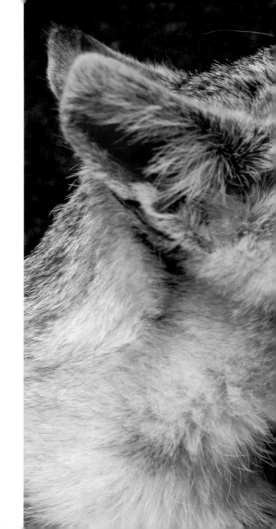

C'mon! Let's arm wrestle!

I can really
sleep all the
better knowing
we have such
tasteful
cushions.

Mmmm, your new shampoo really does smell nice!

Love means
that sometimes
you just have
to be a pillow.

High five!

The
Evolution
of Cat.

Wow, that whisker oil has really worked wonders, you handsome chap!

I love you
mainly for the
smushiness of
your face.

I hope you're not trying to kiss me on a first date?

Unhand me, woman!

Just the thought of moving is making me more tired. Duvet day?

Why can we never just have a nice coupley photo together?

I don't
think this
is what
Dr Seuss
meant!

We know
how cute
we are, and
we're not
ashamed.

I know it looks like we're being lazy, but this is actually a very complicated game of Twister.

Team,
I think
we've been
spotted.

I just feel
really hurt
by what you
said about
my ears.

How can you
sleep when
there are so
many toilet rolls
still to shred?

111

Your eyes are
an ocean I
could swim in
all day.

And when I count to three, attack.

I better not see any mistletoe.

To do: eat, miaow, sleep, sleep, sleep, snuggle.

I want to be
Superman,
but I also
want to sleep.

You have an
uncanny ability
to nap anywhere,
at any time!

It was this
messy when we
got here . . .

We're not that
bright, but we
intend to get
by on our
good looks.

Woah there,
tuna-breath!

I'm sensing a lot of tension in your shoulders.

Handbag?
I thought
you said
ham bag!

Are you
snoring or
purring?

Are we lying on a fluffy rug or is it just a really big cat?

You've definitely definitely got more covers than me!

Hello! I drank an energy drink!